British
FIRE ENGINE
Heritage

British
FIRE ENGINE
Heritage

ROGER PENNINGTON

CHANCELLOR
PRESS

Dedication

For family and friends

First published by Osprey in 1994
This edition published in 1997
by Chancellor Press, an imprint of
Reed Books Limited
Michelin House, 81 Fulham Road
London SW3 6RB
and Auckland and Melbourne

© Reed Books Limited 1994
© Text and photographs Roger Pennington

ISBN 0 75370 012 3

Printed in China

Project Editor Shaun Barrington
Editor Sally Game
Page design Keith Goodenough/Ward Peacock
Partnership

Acknowledgements

All photographs have been taken on Kodak Professional Ektachrome, a versatile colour reversal film; and acknowledgement is made to Colortech Laboratories of Worthing, Sussex, for constant and reliable E6 processing over the years.

There are many, many, people to be thanked for helping me set up and obtain the photographs which have been reproduced, and my sincere thanks go out to all those good friends who have so enthusiastically given time over the last ten years or so to help me. One person to whom very special thanks must be given is John Tassell, Senior Administrator at West Sussex Fire Brigade Headquarters, Chichester, for it is he who suggested that I photograph fire appliances and produce postcards to be sold in aid of The Fire Service National Benevolent Fund. This set of postcards now numbers 55; my thanks to the many collectors who eagerly await the printing of new cards.

Thanks to Mrs M. M. Rowe, County Archivist, City of Exeter; and my good friend Roy Rice who, with his brother Walter, manages a collection of fire appliances in the old Victorian fire station at Banwell, Somerset, who assisted with my historical research. Thanks to Joan my wife, who patiently typed and re-typed the manuscript.

Introduction

The history of fire-fighting equipment, from leather fire buckets dating back to the Roman Empire to today's sophisticated vehicles costing many thousands of pounds, is an intriguing and eventful one. On the following pages I have endeavoured to illustrate this progress with photographs highlighting the changes in fire-fighting technology over nearly 300 years.

Over the last three centuries the design of these appliances has become increasingly complex and diverse. In this one volume it is impossible to show all the many types and styles of fire appliances which luckily remain preserved, along with all those 'on the run' today. I therefore trust you will understand if your particular favourite appliance is not featured in the selection included here.

1910 HALLEY 60/90 HP MOTOR FIRE ENGINE. *A truly handsome piece of work, this Halley fire engine sports a shiny aluminium engine bonnet, polished natural wood, large brass headlamps and distinctive decoration. It was supplied to Leith Fire Department in July 1911, and is now displayed in the Edinburgh Museum of Fire. Powered by a six-cylinder, 60-hp petrol engine, this Halley design was capable of speeds up to 45 mph. It was equipped with a 40-gallon first-aid water tank with provision for carrying a 40-foot telescopic ladder, 1,500 feet of delivery hose and 12 men. The four-forward/one-reverse gearbox followed the 'Halley foolproof pattern'; the road wheels were the artillery type with solid rubber tyres. Manufactured by Drysdale and Company Ltd of Yoxford, Glasgow, the fire pump, known as 'Bon-Accord', was of centrifugal design and capable of delivering 450 gpm at 150 lbs psi. In March 1910 the tendered price for the fire engine was £1,075. Halley Industrial Motors of Glasgow, the manufacturer, was taken over by Albion in 1935, who were in turn taken over by Leyland Motors*

About the author/photographer

Roger Pennington joined Independent Television as a TV cameraman in 1955, and moved on to be Assistant Film Director for many programmes, commercials, features and documentary films on a great variety of subjects. After working for seven years as a film editor, mainly for BBC Television, he moved to directing documentary and travel films for cinema, television and industry. During this time Roger has visited 87 countries, filming everywhere from deserts and coal mines to Royal Palaces. He has also worked extensively around the world as a stills photographer.

Married with three children and three grandchildren, Roger now lives in East Sussex. One of his first childhood memories is of building a 'fire engine' with a garden ladder plus bits and pieces taken from the shed of his parents' home at Cheam Village. He developed a more serious interest in fire appliances in the early 1970s, and now owns and rallies an ex-Royal Navy Rescue Fire Appliance.

Contents

1953 THE SELF-PROPELLED PUMP. *Commonly known as the 'Green Goddess', this pump was designed for strength, high performance and versatility, requiring a minimum of skilled maintenance. It was produced from 1953-56 by several coach-building firms, built exclusively on a Bedford chassis. Ordered and owned by the Home Office, and crewed by AFS personnel, each carried the inscription: 'This vehicle is the property of the Home Office'. The vehicles, measuring 23ft in length and 7ft 4in in width, were driven by a Bedford six-cylinder 4.927 cc 110 bhp petrol engine. The earlier models were fitted with 2-wheel drive and all later models with 4-wheel drive. The Army Fire Service purchased a number of the 4-wheel drive version, which were painted red and known as the Bedford RLHZ. The 2-wheel drive models carried 400 gallons (1,818 litres) of water; the 4-wheel carried 300 gallons (1,363 litres). All models were fitted with a single-stage Centrifugal Sigmund FN4/5 900-gpm rear-mounted pump; a 5½in suction inlet; four standard delivery valves; 6in flange for relay pumping; a 1,600' x 2¾" delivery hose; a 180' hose reel on each side; a light, portable pump; and a 35' aluminium alloy extension ladder. This fully equipped 1953 2-wheel drive appliance is now permanently preserved. At the time of writing 1,100 are still being stored on behalf of the Home Office*

From Buckets and Squirts to Manual Pumps

Much has been written about the early days of fire-fighting, but little detail is known about some of the very early fire appliances, many of which have disappeared into oblivion, together with reliable records of their existence.

History records the formation of fire brigades in the City of Rome as early as 6 BC. These were made up of teams of men known as vigiles, who also acted as policemen. Later they were equipped with large brass syringes, known as 'squirts', operated by a crew of three – two to hold the squirt and the third to work the plunger. However, most relied upon were teams of volunteers passing buckets of water from the source to the fire.

Whilst Britain was part of the Roman Empire, it was known to have its own vigiles, presumably part of the Roman Army centred on the principal settlements. However, with the withdrawal of the Romans from Britain in AD 410, organised fire cover was to disappear.

The next period of recorded British fire-fighting begins with the passing of a law by King Alfred (AD 871-901) stating that all householders were to cover their fires at nightfall in order to reduce the chances of fire breaking out during the hours of darkness.

In Norman times William the Conqueror (1066-1087) introduced a nightly curfew bell which was rung to remind inhabitants to cover their fires. This was known as 'Couvre feu,' hence the word 'curfew'. Henry I (1100-1135) repealed this law due to its unpopularity. However, a requirement for all householders to keep a full bucket of water next to the fire remained. In later centuries larger cities and towns used watchmen to warn of fire, although the most effective hope of extinguishing a fire once it had taken hold was a downpour of heavy rain or a change in wind direction.

Early written reports recall a fire engine seen in the City of London's Lord Mayor's Show in 1548. Presumably, it was mounted on a form of sledge. Later, Patent Office records in London indicate that the first patent for a fire appliance was granted to a Roger Jones in 1625, describing it as a 'fire extinguishing machine...with a spout of copper or

1710 LEATHER FIRE BUCKET.
Dating back to the Romans, the bucket remained an essential piece of fire-fighting equipment for many centuries. This original riveted leather fire bucket, made in 1710, has a brass emblem of The Sun Insurance Company. As a result of the Great Fire of London in 1666, insurance companies were to become responsible for forming their own fire brigades, with Sun Insurance claiming to be the first in 1710. Insurance companies handed out dozens of leather buckets to their insured clients for distribution in their buildings – bearing the company crests, the buckets served as free advertising

1728 MANUAL FIRE PUMP. *The earliest known example of a Richard Newsham design, this is believed to be the oldest complete fire appliance preserved in Great Britain. Purchased new from London in 1728 by the Churchwardens of St. Giles Parish Church, Great Wishford, Wiltshire, the appliance cost £35 delivered. It is still displayed at the Parish Church, and contains the original operating instructions, including advice for the watering of domestic gardens when not required for fighting fires! (With acknowledgements to the Churchwardens of St. Giles)*

brass'. The patent describes its use and components in great detail, and boasts that the machine 'will quench the fire with more ease and speed than 500 men with the help of buckets and ladders'. Roger Jones only built two machines, and his patent ran out after 14 years.

Reference to a fire engine has been found in early City of Exeter records. The minutes of a meeting in 1652 tell of the appointment of Mr Henry Prigg to write to London about obtaining an 'engine or spowte for the quenching of fire', and the City Accounts record a payment of six shillings to Mr Prigg for his expenses incurred in the carriage of the 'engyn and bucketts' from London. However, no record of payment for the fire engine can be found.

In a programme from an International Fire Exhibition held at Earls Court, London in 1903, there is mention of an exhibit from the Exeter Fire Brigade, described as a two-man manual engine from 1626. 'In case of fire, application for its use had to be made to the Mayor or Beadle,

who would loan the engine together with the buckets, on condition that the whole was returned in good repair; the engine being carried to the fire by means of hand poles and shoulder straps'. It states that the engine was supplied to Exeter in 1626 and was the sole protection of that City for upwards of one hundred years. (It is assumed that this engine is the one referred to at the Exeter Records Office.)

It was the Great Fire of London on Sunday, 2nd September 1666, that resulted in fire-fighters being formally organised and serious thought given to the design and manufacture of fire-fighting equipment. This disastrous fire, which burnt for five days, destroyed 13,200 homes and 87 churches, including the original St Paul's Cathedral.

Due to insufficient fire-fighting equipment at that time, the flames spread uncontrolled and eventually reached the edges of the City. Explosives were used on the orders of King Charles II (1660-1685) to produce fire breaks to prevent the destruction of the rest of London.

1760 EARLY MANUAL WITH SPOUT. *It is believed that this pump is a second size Newsham pump, built by Newsham and Ragg. Ragg, the nephew of Richard Newsham, ran the firm after Richard Newsham, his wife, and their son Lawrence had all died. According to a Newsham advertisement, the appliance could hold 36 gallons of water, discharge at the same rate per minute, and reach 84 feet through the spout pipe directed by a fire-fighter standing on top of the pump*

Altogether over three quarters of the buildings in the City of London were destroyed. Amazingly, only six people were killed. Damage was estimated at over £10 million for buildings alone.

As a result of this monumental fire, insurance companies were to become responsible for forming their own fire brigades, with Sun Insurance claiming to be the first in 1710. Soon afterwards other insurance companies followed suit, each forming their own brigade. Equipment consisted of a handcart containing a short ladder, leather buckets, and maybe a couple of heavy hand squirts, which when used in succession, gave a continuous flow. At the same time, insurance companies handed out dozens of leather buckets to their insured clients for distribution in their buildings. Bearing the various insurance companies' crests, the buckets acted as free advertising.

During the early 17th century, it appears that Germany led in the design and manufacture of several types of fire pumps. In 1688 Dutch fire

Above
1820-1853 MANUALLY-OPERATED FIRE APPLIANCE. *This appliance, manufactured by Hadley Simkin and Lott, was based on a design by Richard Newsham, although the King's Patent was granted to the former. Purchased by the Duke of Devonshire for the Parish of East of Bourne, now better known as the south coast resort of Eastbourne, it was replaced in 1853 by a steam appliance, paid for by public subscription. The appliance shown is now preserved at Eastbourne Fire Station*

Left
1809 MANUAL FIRE PUMP. *Manufactured by Hadley and Simpkin, Long Acre, London (later to become Merryweather & Sons, Ltd), this pump was built under the 'King's Patent' and was a signficant improvement over earlier manual types. In order to increase the normal water delivery which was pumped from the water trough, a second crew could tread 'in time' holding on to the upper handrails. The trough could be filled either manually or by leather suction hose. The pump shown here served the village of Okeford Fitzpaine, Dorset, and is now in the care of the village museum*

appliance manufacturers exported various products to Britain, which were very quickly copied. (A British Patent for one such product was granted to a John Loftingh in 1690.) Sewn leather delivery hose was introduced from Holland, which enabled water to be directed to the seat of the fire. This was later superseded by copper-riveted leather hose, due to eventual rotting of the stitching after continuous use.

Only one 17th-century appliance is known to exist today. It is incomplete, and now resembles a highly polished, open-top wine cask rather than a piece of manual fire equipment. Built by John Keeling of London, it was sold to the town of Dunstable in the 1670s and is now displayed in the City of London Museum.

After three quarters of a century of various manual pump inventions, none of which appear to have been commercially successful, it was a pearl button maker from Smithfield, London who finally got it right. In 1721, Richard Newsham was granted a patent for his revolutionary fire

1850 MERRYWEATHER HOSE AND LADDER CART. *A typical example of those used by the early insurance fire brigades, such handcarts have a short extension ladder to gain access to first-floor windows, a dozen or so leather buckets, and one or two brass squirts. First supplied to Crieff Hydropathic Hotel, this handcart can now be seen in the Edinburgh Museum of Fire. Later models, used until the turn of the 20th century, carried a hydrant stand pipe and lengths of canvas hose. In a notice dated July 1913, the Baldock Fire Brigade announced a charge of two pounds and two shillings per 24 hours work for a handcart of the type shown*

pump, which could throw 110 gallons of water per minute.

His design was based on a long forgotten, 2,000-year-old idea, which used an air chamber to equalise the output pressure of the pump. Overnight he had become a successful engineer, manufacturing several different sizes of pump, and just four years later he was granted a second patent. It is to his credit that Newsham Pumps are still to be found today in working order in museums and private collections. One of his earliest known pumps (which is believed to have successfully fought a fire in the 1920s) can be seen in the Parish Church of Great Wishford, near Salisbury, Wiltshire. When purchased by the Church Wardens in 1728, this pump cost just £32.3s.0d. plus delivery.

Without a doubt manual fire pumps had arrived by 1740, and so had Newsham's competitors. However, his pumps remained popular with both British and overseas customers, and production of six different models continued for nearly 100 years.

Upon his death in 1743, the business passed to his cousin, George Ragg. The Company then traded under the name Newsham and Ragg until at least 1765.

Two more pump manufacturers were to emerge – John Bristow and Charles Simpkin of Longacre, London. Simpkin was eventually to merge with Merryweather, an established fire engineering manufacturer, which began trading in 1692 producing leather fire buckets.

Merryweather was to become a household name, supplying country homes with domestic fire-fighting equipment and fire brigades with larger

1860 MERRYWEATHER 'TOZER' PUMP. *This appliance was invented by Alfred Tozer who was chief officer of the Manchester Fire Brigade from 1861. It was still being advertised in the Merryweather catalogue during the first quarter of this century, described as follows: 'This engine has double the power of the London Brigade Fire pump, being capable of discharging 12 gallons of water per minute to a height of 45 feet; it is fitted with a brass rail round the top to enable two persons to carry it when full of water...The Pail is handsomely painted vermilion, or any colour to order, and tastefully finished, thus combining utility with elegance...Contents of pail seven gallons.....Price complete with 10 feet of leather hose, screws and jet £7.0s.0d.'*

1859 MERRYWEATHER MANUAL FIRE APPLIANCE. *This Merryweather pump, capable of delivering approximately 55 gpm, was worked by four men on each side. In a Merryweather handbill produced for the Great Exhibition in 1851, this model was described as a 'Portable Fire Engine for Ships, Entrance Halls of Mansions, Club Houses, Banking Houses, etc.' This particular appliance was manufactured in 1859, though the design dates from about 1840. It has been restored by members of Felixstowe Fire Station, where it is now displayed*

1877 MERRYWEATHER MANUAL PUMP. *This pump was purchased second-hand in 1883 by J.H.C. Evelyn, Esq. of Wotton House, Nr. Dorking, Surrey. (Wotton House was used for many years as the Fire Service Staff College). With a 6-inch Merryweather engine, it was designed to be worked by 30 men, producing 100 gpm at one stroke per second. The pump was drawn by two horses, and weighed 1½ tons when loaded. Found in a derelict condition, it was restored by firemen at Dorking Fire Station, and reportedly can still produce a good fire-fighting jet*

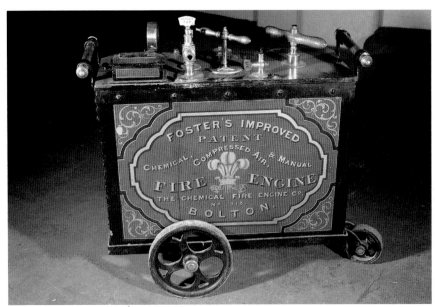

appliances. They eventually ceased trading in the mid-1980s.

The next half-century saw various changes to the design of these basic appliances, including the addition of wooden shafts which enabled the pump to be drawn by horses. For the next major leap forward, fire and water would become allies, rather than enemies.

Above

1880 FOSTERS IMPROVED FIRE ENGINE. *Fosters of Bolton described their unusual and 'improved' product as a chemical, compressed air and manual fire engine. Fire engine no 413, pictured here, was purchased by Sandringham Fire Brigade in 1882 at a cost of £21.15s.0d, for use on the Sandringham Estate. It can now be seen in the Fire Brigade exhibition in the Museum at Sandringham, Norfolk, the present private country retreat of the Royal Family. (Photographed by gracious permission of Her Majesty The Queen)*

Left

1880 SHAND MASON. *Little is known about this appliance. It is thought to have served somewhere in London when new, and eventually at Bishops Stortford, where it was found in a derelict condition in the 1970s. Ownership eventually passed to the National Museum of Fire Fighting, and a benefactor recently paid for the restoration of the wheels and paintwork*

From Steam to Motorized Vehicles

Europe's first steam fire appliance was built in Britain in 1829 by Braithwaite and Ericsson of London. Although only four more steam appliances were to be built in the next four years, the design was to be an important milestone.

Despite this invention, production of manual pumps continued into the 1860s. Merryweather, Shand Mason and a lesser known company – Baddley and Roberts of London – sold many of the pumps to factories, hospitals and country estates (where the Squire of the Manor would elect himself Fire Chief).

It was another 29 years before Shand Mason produced their first steam pump in 1858, Merryweather following with their first in 1861. In 1862 an international exhibition of mechanical inventions held in London's Hyde Park included two Shand Mason steam appliances, and one by Merryweather; all three caused great interest. One year later the Merryweather pump was to be awarded first prize in the Heavy Pump Class at a trial held at Crystal Palace, London, with seven pumps from Britain and three from America competing. Shand Mason was awarded first prize in the Small Pump Class. These awards led many British Brigades to place orders for this type of appliance.

In 1900 Merryweather took the next logical step by producing a self-propelled steamer, 'The Fire King'. Within six years the company had manufactured 30 of these steamers for both overseas and home markets.

With the arrival and advancement of the petrol-driven motor car, motorized fire appliances soon took over from the horse-drawn steam appliance and the small number of steam-propelled machines.

The first two attempts at producing motorized appliances turned out to be badly calculated failures, due to underpowered engines, but in 1903 Merryweather, now a well-established company supplying fire-fighting

1893 SHAND MASON HORSE-DRAWN STEAM FIRE-PUMP. *The No. 1 Shand Mason vertical-type pump delivered a 1¾" jet, 160' high, at a rate of 350 gpm. It was built in 1893 and purchased the same year by Arundel Castle, Sussex, for £450. In 1949 it was presented by the late His Grace the Duke of Norfolk to the West Sussex Fire Brigade, and it was later completely renovated and rebuilt by firemen from Worthing Fire Station. It was a custom for steam pumps to be given a name, usually the name of the Castle, County Manor or village where it was kept*

1908 SHAND MASON STEAM FIRE PUMP. *Lord Shaftesbury asked Shand Mason to build this steam pump for the protection of St. Giles House and Wimborne St. Giles village in Dorset, after the village's manual pump proved inadequate to combat the fire which destroyed the parish church in 1907. The steam pump faithfully served the village until 1948. Now preserved in full working order, it appears regularly at county shows and fire rallies*

equipment both at home and overseas, succeeded with a highly-acclaimed motor-driven appliance – the first to be supplied to a public fire brigade.

The next year, Merryweather manufactured another first-of-its-kind appliance, this one with a Hatfield 500-gpm pump driven directly by shaft from the road engine. Fitted with a 50-foot wheeled escape and a 60-gallon first-aid water tank, it served London's Finchley Fire Brigade for 24 years until 1928. The chassis, engine and pump of this appliance can be seen in the Fire Fighting Gallery of the Science Museum, South Kensington, London.

New companies soon entered the marketplace, but few survived. Dennis Brothers of Guildford were an exception, along with Merryweather, who were still producing their successful Hatfield Pump appliance. In 1910 Merryweather made the change to an Albion chassis, which proved to be very successful.

At the same time, in 1910, a short-lived company in Scotland called Halley Industrial Motors offered a range of fire appliances known as `Halleys'. One of these can be viewed in the appliance bays at the Old City Fire Station Museum, Edinburgh (just behind the Castle), now the headquarters of Midlothian and Borders Fire Brigade.

Another well-known chassis producer was Leyland Motors Limited of Leyland, Lancashire, which began fire appliance production in 1907, and delivered their first appliance to the Dublin Fire Department in 1910.

With the successful development of motorized appliances, horses were soon to be phased out. For the firemen, the love and care lavished on their horses was to become a cherished memory, and for the public, the Saturday morning treat of visiting the fire station stables was to become a thing of the past.

1914 DENNIS BRAIDWOOD PUMP. *Powered by a White and Pope 9.1-litre, four-cylinder, side-valve petrol engine, this pump is still fully operational. It was originally supplied to the City of Coventry in May 1914, and was the first to replace the city's horse-drawn appliances. After servicing the city for 20 years, the pump was sold to G.E.C., a local electrical engineering works, who retained it as a factory appliance until 1958. During this period the original solid rubber types were changed to pneumatic ones. Dennis Brothers, its builders, then purchased the pump for £35, and returned it to their Guildford Works, where it was looked after by apprentices. Today it is maintained by the Dennis 1914 Association of Godalming*

1912 BELSIZE / JOHN MORRIS PUMP ESCAPE. *Originally supplied to the Southampton Fire Brigade in 1912, this pump escape was named 'Madeleine' after the wife of the Mayor. Having been driven from the John Morris Works in Manchester under its own power, it required a new set of rubber tyres for its wooden wheels upon arrival in Southampton. The pump escape was sold by the Southampton Fire Brigade in 1926 to a Works Brigade, who kept it until the late 1950s. In 1961 it was found in a scrapyard and bought by The Enfield and District Veteran Vehicle Society for £50. Equipped with 50' John Morris wheeled escape, it is powered by a 6 cylinder Belsize 14-litre petrol engine, with 5¾-inch stroke, 4 forward gears and separate lever for reverse, and open chain final drive*

1934 DENNIS BRAIDWOOD BIG 4 PUMP ESCAPE. *A two-time winner of the Royal Automobile Club's London-Paris-London race, this reliable appliance has another claim to fame: in 1980, crewed by men from The London Fire Brigade, it was driven from John O'Groats to Land's End in eight days to raise money for the Fire Services National Benevolent Fund. Supplied in 1934 to Letchworth Garden City Fire Brigade in its distinctive green livery, the 3 1/2 ton, 10-man machine is powered by a Dennis D3 four-cylinder engine, and fitted with a No. 3 Dennis main pump, a 2,000' hose and a 50' Merryweather wheeled escape. It served Hertfordshire Fire Brigade as a reserve appliance from the mid-50s until 1963 when it was sold for preservation and restored to its original specification*

1934 DENNIS 30 CWT OPEN BODY. *This handsome appliance, complete with 'Avon' tyres, Ash bodywork and framework (including special panels), was fitted with a Dennis No. 2 pump, three 8' x 31/2" suction hoses, 1,000 feet of delivery hose and a 30' telescopic ladder. It was built on Dennis 30 cwt chassis no. 7663, and powered by a White and Pope four-cylinder 85 x 120 DHV engine. Supplied new in August 1934 to Avon India Rubber Co. Ltd. in Melksham, Wiltshire, for the cost of £667 and five shillings, it was used during the Second World War at Southampton Docks. The appliance was sold at auction in 1986 for £8,100*

1935 LEYLAND 'CUB' MOTOR PUMP. First introduced in late 1935 as the FK6, this pump was known for its 'Braidwood' body design (in which the crew sat facing outwards). Its six-cylinder Leyland 29.3 hp petrol engine had a power output of 62 bhp. Other features include an 11'6" wheelbase, a 4-speed gearbox and a rear-mounted 350-500-gpm Dennis pump, although some appliances were equipped with a 500-gpm Gwynne pump. (Type FK7 was supplied with its pump mounted amidships.) This particular 'Cub' pump was purchased in 1940 by the Plympton Rural District Council, Devon, and was stationed at Plymstock until 1964, serving in Plymouth during the 1941-43 blitz. In 1948 it was transferred to the Devon County Fire Service

Above

1935 DENNIS BRAIDWOOD OPEN-BODIED 'BIG FOUR'. *Originally supplied in green livery to Bedford Fire Service in 1935, this 3 ton 'Big Four' featured a D3 four-cylinder side-valve engine fitted with dual ignition. Built with seating arrangements for 10 men, it is equipped with a 40-gallon first-aid water tank, a Dennis No. 3 pump, a 180', ¾" hose reel, lockers to hold 2,000 feet of hose and a 50' Bayley wheeled escape. The appliance is now in preservation*

Right

1936 DENNIS 'ACE' PUMP. *Powered by a Dennis 3770cc side-valve petrol engine, and equipped with a No. 2 Dennis fire pump capable of delivering 500 gpm, this appliance is still operational today. It has been the Works appliance at the Dennis factory in Guildford since it was built in 1936, and is still looked after by the young technicians at the factory*

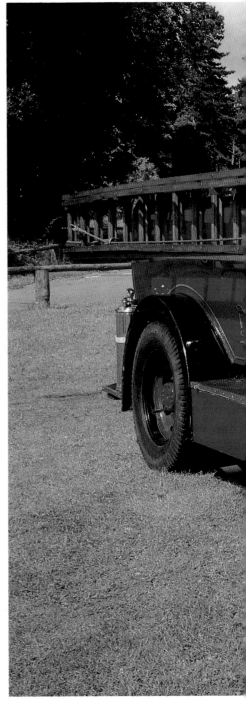

Above

1937 LEYLAND 'LYNX' ESCAPE/HOSE REEL TENDER. *Bought new by the Skegness Urban District Council, the 'Lynx' faithfully served the seaside town until the 1950s. In 1939 it was driven to the Dennis factory at Guildford to be fitted with a Dennis No. 2, 500-gpm pump, and a Merryweather 50' all-metal wheeled escape, thereby converting the appliance to a pump/escape. (It is interesting to note that the Lynx is not listed by Leyland, although the appliance's original papers all record it as such, adding another name to the Leyland animal kingdom of Cub, Tiger, Lioness, Terrier and Beaver.) This photograph was taken on Madeira Drive, Brighton, at the end of the 1993 Historic Commercial Vehicle Run from London*

Left

1938 DENNIS LIGHT SIX-PUMP ESCAPE. *Also known as the 'Big Ace', this model was first produced in 1936 and remained in production until 1939. This particular appliance was supplied new in 1938 to the County Borough of Reading Fire Brigade, remained 'on the run' at Caversham Road Fire Station until 1947, and was then transferred to the new Berkshire Brigade as a reserve vehicle. Sold as scrap in the 1960s, it was restored by Sir Nicholas Williamson of Mortimer and handed over to the Berkshire Veteran Fire Engine Society for preservation. The appliance is powered by a six-cylinder Meadows petrol engine and equipped with a Dennis 500-gpm pump and a 50' wheeled escape*

1938 LEYLAND PUMP ESCAPE. *This was one of the first enclosed Limousine-style fire appliances, and is the only one remaining of the six built in 1938. It was supplied new to Poole District Council at a cost of £1,758.10 shillings, was transferred to Weymouth, Dorset in 1953, and remained in service until the mid-1960s. Fitted with a Leyland six-cylinder petrol engine and a Leyland two-stage turbine-type pump, the appliance has bodywork by Leyland*

1938 LEYLAND CUB FK6 OPEN HALF CAB. *First introduced in 1936 using a Braidwood-type body, this appliance was supplied new to Windermere Rural District Council in 1938, condemned by the National Fire Service in 1941, and transferred to Liverpool Docks for the duration of the Second World War. The FK6 is powered by a Leyland type E 109-3 six-cylinder petrol engine, and fitted with a 500-gpm Gwynne Centrifugal rear-mounted pump and a 100-gallon first-aid water tank. Purchased for preservation in 1977, it has been beautifully restored to its original condition, as shown in this photograph taken at the 1984 Blandford Fire Engine Rally*

Above
1938 DENNIS NO. 2 TRAILER PUMP.

Left
1938 LEYLAND 'CUB' SEMI-LIMOUSINE FK6 FIRE APPLIANCE. *The Caerphilly Urban District Council purchased this appliance in 1938 and, during the Second World War, loaned it to the London County Council. In 1946 it was returned to Caerphilly and remained in service with the Glamorgan Fire Brigade until the late 1950s. During the early 1960s it was used as a 'Hearse of Honour' for Brigade funerals, and in 1968, after being found in a derelict condition, it was rebuilt and appeared in the film* The Battle of Britain. *The appliance is powered by a six-cylinder 29.4 hp OHV engine (petrol consumption being 6.7 mpg) and is equipped with a 500-gpm Gwynne main pump*

Above

1939 MERRYWEATHER 'HATFIELD' MOTOR FIRE ENGINE. *King George VI bought this handsome 'Hatfield' to replace an earlier Merryweather appliance which had served Sandringham Fire Brigade for 23 years. It is believed that the 400-gpm Hatfield pump from the original appliance was transferred to the new one. It was loaned to the City of Norwich to fight fires from enemy bombing in 1942. It was eventually returned to its home and is now featured at an exhibition in the original Fire Station at Sandringham, Norfolk. Featuring a Braithwaite-style body built of seasoned mahogany, this showpiece is equipped with a 'Telescopic' extension ladder, a powerful searchlight and a six-cylinder 60 hp petrol engine*
(Photographed by gracious permission of Her Majesty The Queen)

Right

1939 DENNIS 'LIGHT FOUR'. *Built to carry a crew of four, this unusual design features an enclosed cab on a Braidwood body and measures 19' in length by 8'8" in height. Only six such appliances are believed to have been built. A four-cylinder Dennis petrol engine powered the 'Light Four', and it was fitted with a five-forward/one-reverse gearbox. Equipment onboard included a Dennis No. 2 500-gpm rear-mounted pump, a 35-gallon water tank, one hose reel, a 35' Ajax ladder, and two hook ladders. First supplied to Watford Rural District Council in 1939, this appliance served with Hertfordshire Fire Brigade from 1946 to 1957, and was then sold to Works Brigade, Murphy Chemical Co., Wheathampstead*

1939 DENNIS 'NEW WORLD LIGHT FOUR'. *Built on a Dennis 'Ace' chassis, the New World Light Four body afforded greater protection to the crew members who sat facing inwards, instead of outwards as on the earlier Braidwood design. It is powered by a four-cylinder engine, fitted with a type 'P' series 37 gearbox; and equipped with a Dennis No. 2 375-500-gpm.centrifugal pump, a 35-gallon first-aid water tank, a 120', ¾" hose reel with central branch, and a 30' Ajax ladder. The 3 ton truck, which could carry a crew of eight, was first supplied to Gipping (Suffolk) Rural District Council in 1939 for Needham Market. It was transferred to the National Fire Service in 1942. In 1966 the Felixstowe Dock and Railway Company purchased it as a back-up appliance for use at Felixstowe Docks*

Right

1940 ALBION 127 OPEN-CAB FIRE APPLIANCE. *Known for its fine collection of brass fittings, this open-cab appliance was built with a 27.6 hp engine for Pilkington Bros., Glass Manufacturers of Doncaster. It was sold to Rainhill Hospital, Merseyside in 1968, and was finally purchased for preservation in 1982*

The Second World War

With the threat of European war in 1938, the British Government had already begun to organise fire-fighting manpower and equipment.

The Auxiliary Fire Service, better known as the AFS, was formed, with the Secretary of State at the Home Office taking overall charge of some 1,500 large and small fire brigades scattered across Great Britain. Due to a range of varying appliances and equipment in each brigade, plans were made to supply standard fittings and connections, to ensure equipment would be compatible in the event of an emergency.

By the time war was declared with Germany on 3rd September 1939, contracts had been issued for the construction of hundreds of purpose-designed auxiliary towing vehicles, based on the Austin K2. The Austin Motor Company were awarded the largest contract for these appliances, known as ATVs or K2s. When delivered, these K2s replaced the hundreds of commandeered taxis, vans, trucks and other assorted vehicles which had been used to tow trailer pumps, many of which were far too heavy for these temporary towing units.

During the course of the war some 9,000 ATVs were built. Today, 40 or so remain in preservation, with the rest now broken up or serving in some remote part of Britain as a wood or chicken shed.

Later the Ford Motor Company were also to produce ATVs on their War Office type chassis, a more robust vehicle than the Austin K2.

To equip the newly formed AFS units and other industrial users, many hundreds of trailer pumps were ordered by the Home Office. These ranged from a No. 2 Dennis 500-gpm trailer-mounted pump, to the well-known Coventry Climax and the lesser-known Stork utility trailer pump, which had an output of 120 gpm, being powered by the famous Austin 'Seven' four-cylinder petrol engine.

With the supply of Metz and Magirus turntable ladders from Germany no longer available, another well-known wartime appliance was to

1940 DENNIS LIGHT 4 PUMP ESCAPE. *A four-cylinder Dennis side-valve 3769cc engine drove this 1940 appliance. Fitted with a 5-speed gearbox, it carried a Dennis No. 2 500-gpm rear-mounted turbine pump, a 35-gallon first-aid water tank and a 50' wheeled escape ladder. The appliance was supplied new to Hertford Corporation in February 1940, and was purchased for preservation in 1953*

THE ROYAL TO
A12 CIVIL-DEFENCE
A12. A.R.P. STREET-FI
AIRRAID-WARNING
TAKE
COVER
CNO 871
G R

Above

1938 COVENTRY CLIMAX FSM-TYPE LIGHT-TRAILER FIRE PUMP. *This 214-gpm Godiva single-stage centrifugal pump was driven by a Coventry Climax four-stroke, four-cylinder, water-cooled petrol engine. Designed in conjunction with the Home Office, many hundreds of these trailer pumps were built at the outbreak of the Second World War, and were used for first attendance at Air Raid fires. Although widely rejected because of their limited output, many were stored by the Home Office after the end of the war*

Left

1935 AUSTIN LITCHFIELD A.R.P. WARDEN STREET FIREPARTY CAR. *Although some might not consider it a fire appliance, this rare car was converted in 1941 by the members of the A 12 Air Raid Wardens and Firefighters Post at Clacton-on-Sea, Essex. It carried an assortment of hand equipment, including a dustbin lid for extinguishing incendiary fire bombs. In 1990 the 'fireparty car' appeared in The Royal Tournament held in London*

emerge – the Austin K4 hand-operated, three-section Merryweather 60-foot turntable ladder unit. Only a handful of these have been preserved.

To help supply water to major fires caused by heavy bombing, where on many occasions water mains had also been damaged, two other war-time vehicles were developed: the heavy pump unit, which was designed to pump water in relays to bombed areas (on one occasion water was pumped through nine miles of hose); and the dam unit, which was a portable canvas dam supported by a metal frame placed on the back of a flat-bed lorry.

Above
1940 FORDSON 7V HEAVY PUMP UNIT. *Issued new to Aberdeen Auxiliary Fire Service in 1940, this heavy pump unit was transferred to the National Fire Service in 1941. After the war it served the North Eastern Fire Brigade until the mid 1950s, when it was bought by the local paper mill. Upon closure of the mill, it lay in the open until 1988 when it was purchased for permanent preservation*

Right
1942 AUSTIN K2 AUXILIARY TOWING VEHICLE. *Reputed to be very reliable and versatile, the K2 was ordered by the Home Office in large numbers in the early years of the Second World War to tow heavy trailer pumps. Built on an Austin 2 ton chassis with a capacity for six crew members, the interior was fitted with bench-type seats, the underneath area being used for storage of hoses and equipment. A 30' ladder was carried on the roof. The GLT 396 shown here was rebuilt by Mick Paull, taking over 600 hours of painstaking restoration work. The pump being towed is a Dennis TP 500-gpm*

After the cessation of hostilities, the National Fire Service, which had been formed in August 1941 by amalgamating local fire brigades and the newly formed AFS, was disbanded, and responsibility for fire fighters once again reverted to County and Local Authorities.

Many of the wartime machines, which had been painted a uniform grey, were transferred by the Home Office to the newly formed fire brigades. Overhauled and repainted in conventional fire-engine red, many remained in service until the mid-1950s.

Above

1942 AUSTIN MULTI PURPOSE PUMP. *Mick Paull purchased this Austin pump for preservation in 1979, and spent 900 hours restoring it to its original condition. First supplied in 1942 to the National Fire Service in Berkshire, the pump was stationed at Hungerford for several years where it carried the name 'John O'Gaunt 2' (the name being transferred from the previous appliance which was originally named after a local benefactor.) After the disbandment of the National Fire Service in 1948, the appliance passed to the Berkshire and Reading Fire Brigade. It was sold in 1959 to Adwest Engineering at Woodley Aerodrome, to be used as a works appliance, and was subsequently burnt away on one side when the wind suddenly changed whilst fighting a fire, making the restoration task a daunting one*

Right

1943 AUSTIN K4 / MERRYWEATHER 60' ESCAPE. *Demand for this unique utility appliance soared during the Second World War. A total of fifty were built at the Austin Motor Co. Ltd Works in Longbridge, Birmingham, including the fine and well-preserved example pictured here. Powered by a 3.5 litre, six-cylinder Austin petrol engine, the 5 ton vehicle features a four-speed gearbox and four-wheel hydraulic brakes, along with a three-section manually-operated 60' Merryweather Turnable Ladder. Many of these vehicles had a front-mounted 350-gpm fire pump as well*

Above
1943 AUSTIN K4 ESCAPE CARRIER UNIT. *This well-preserved, prize-winning appliance was first supplied in grey livery by the Home Office to Pembroke Docks in South Wales in 1943. After the Second World War it was taken over by Pembrokeshire Fire Service and stationed at Tenby, and years later it was briefly housed at a museum in Milford Haven. It is now owned and maintained by Ted Pither of Odiham. Fitted with a front-mounted Barton Pump, the appliance has been beautifully preserved in traditional red livery*

Right
1943 FORD 'HEAVY PUMPING UNIT'. *Many of these units were built during the Second World War and supplied to the National Fire Service by the Home Office. This type of appliance was designed to pump large amounts of water from open water supplies to the scene of large fires. The one seen here was built in 1943 by the Ford Motor Company on a Ford chassis, known as a Model 7V, and powered by a Ford 85-bhp V8 engine with a four-speed gearbox. Weighing in at 4¼ tons fully equipped, the vehicle carried a 30' ladder and a crew of six. This appliance is of particular interest as it is one of the few remaining still fitted with the original self-contained 'Tangye' 700-gpm pump driven by a separate Ford V8 engine. Also noteworthy is the steel sheet over the driving cab, affording extra protection against bomb and shell shrapnel. The unit served with the N.F.S. during the War, until it was handed over to West Sussex Fire Brigade in 1951. It was returned to the Home Office in 1956 and, upon disposal, was bought in 1967 by its present owner for preservation*

Above
1943 FORDSON WAR OFFICE TYPE 2H A.T.V. *After discovering this vehicle up to its axles in mud in a farmyard where it had lain for 23 years, the finder spent three years restoring it to its original condition. Auxiliary Towing Vehicles of this kind were built on a Ford chassis which was used for both military and fire-fighting applications, due to the large number of ATVs required following the heavy air raids during 1941-42*

Left
1943 SCAMMEL ULTRA-LIGHTWEIGHT FIRE PUMP WHEELBARROW TYPE. *'A fire station in a wheelbarrow' was the term used for this unique and efficient pump. It proved to be very popular and was issued extensively to factories engaged on war munitions during the Second World War. Built in 1943 by Scammel Lorries of Watford, it has a capacity of 65 gpm at 50 psi. A similar pump called a 'Sled', mounted on a form of sledge, was issued to the Royal Navy for use in both ships and shore establishments*

1943 DODGE 82A MOBILE DAM UNIT. *Despite lying in an apple orchard for a number of years, this machine is in surprisingly good condition. It is one of many similar mobile dam units urgently built during the war to supply fire-fighting water to blitzed areas. Built on a normal production chassis, the unit is fitted with an open crew shelter behind the driving cab and a standard dam unit fixed to the open truck deck. After the stand-down of the National Fire Service, this vehicle served at Datchet Fire Station, Buckinghamshire*

Fire Boats

By the mid 19th century, cities with major waterways running through their midst were beginning to pose great problems for fire-fighters, due to the increase in developments springing up along the waterfronts. London was one such city, becoming the world's largest port with miles of wharves and bonded warehouses. A salutary lesson had been learned from The Great Fire of London some 200 years before, when riverside buildings were engulfed in flames, fuelled by spirits, eastern oils, spices and tallow.

The first record of a floating fire appliance in London – `an engine quenching fire on board of ships being placed in a boat or loyter' – is traced to around 1715. This type of operation proved to be expensive, for not only were the men hired to pump water paid the enhanced rate of one shilling (5p) for the first hour and sixpence for each hour thereafter, but large quantities of beer were consumed by the crews who refused to pump when the beer ran dry.

By 1833 the city's insurance companies, who each had their own fire brigade, agreed on the formation of a joint brigade which became known as The London Fire Engine Establishment. Soon thereafter, under the supervision of Superintendent James Braidwood, tug-towed pump floats were introduced. In 1855 the first purpose-designed, floating steam fire engine was commissioned, based on a conventional ship's hull. Ironically, it was at the first fire attended by this vessel near London Bridge that James Braidwood was killed by falling masonry.

In 1866, now under the command of Captain Eyre Massey-Shaw, the steam fire pumps were remounted onto rafts because the shallow depths of water prevented access by fire boats.

New fire boats were commissioned at the turn of the 20th century. The first, 'Alpha II', was a coal-fire-driven vessel built at the cost of £6,500, which remained in service for 25 years. The second, `Beta', which cost nearly £11,000 to build, had four Shand Mason pumps capable of discharging 1,000 gpm.

Probably the most famous of all fire boats, 'Massey Shaw' cost £17,000 and was launched on 25th February 1935. Although designed as a fire boat, she was to become a household name in 1940 after her three epic journeys across the English Channel to the French beaches of Dunkirk, where she rescued British troops who had been surrounded by

DUNKIRK PLAQUE. *In May 1940, the Massey Shaw joined the armada of little ships which crossed the English Channel to help evacuate British Forces trapped on the beaches of Dunkirk. One hundred and six soldiers were returned to Ramsgate, plus 40 French seamen who were picked up from a sinking ship. The plaque commemorates this unusual tour of duty*

1935 FIREBOAT 'MASSEY SHAW'.

Commissioned by the London Fire Brigade, this rugged fireboat was launched at Cowes, Isle of Wight, on 25th February 1935. She played a major part fighting riverside fires during the Second World War blitz on London. Measuring 78' overall, 13'6" wide at the beam and drawing just 3'9", she is powered by two eight-cylinder 160-bhp Gleniffer diesel engines, which operate her 3,000-gpm centrifugal fire pumps. The Massey Shaw retired in 1971, after 36 years of fire-fighting on London's River Thames. Eleven years elapsed before she was fully restored to working order by the Massey Shaw and Marine Vessels Preservation Society

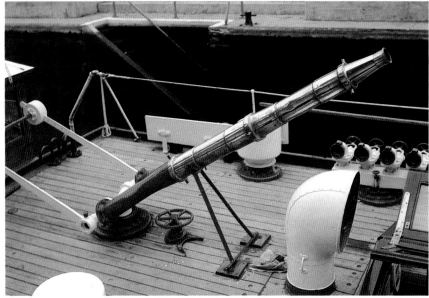

the enemy. In addition, she ferried an estimated 500 other personnel to larger ships waiting offshore, despite bombing and aerial attack. Today, `Massey Shaw' can still be seen cruising the River Thames on special occasions, crewed and preserved by hardworking volunteers.

During the Second Word War, 30 auxiliary fire boats were pressed into service on the River Thames in London, and other ports followed suit. On inland canals traditional 7-foot-wide narrowboats were commandeered and fitted with two portable pumps, such as the Coventry Climax, and held in reserve.

After the War, the older boats in London were phased out of service, partly due to reduction of fire hazards with the decline of the river's commercial waterfront. In the mid 1970s two new boats were built, the `Fire Swift' and the `Fire Hawk', costing £46,000 and £60,000 respectively; and in 1985 `Phoenix' was launched, costing £500,000. In her first two years of service, `Phoenix' answered 1,500 emergency calls.

1975 FIREBOAT 'FIRE HAWK'. *This London Fire Brigade fireboat, built with a 45' glass-reinforced plastic hull, was powered by twin Perkins six-cylinder turbo-charged diesel engines, driving two 24" bronze 3-bladed propellers, giving a maximum speed of 17 knots. It was built in 1975 by Watercraft Ltd. at Shoreham-by-Sea, Sussex. Two Godiva FWBP pumps are fixed to the aft cockpit, with a combined performance of 1,160 gpm at 60 psi, and with quick release fittings to facilitate exchange or derricking aboard ships. The boat's beam measures 13' 5½" (4.1m), and its draught (laden) measures 4'1" (1.25m). Spray jets are fitted below the hull's fender, and the handrail is perforated to direct jets of water over the superstructure to protect the vessel from radiant heat*

Military Equipment

The Fire Service Units of the Army, the Navy and the Royal Air Force have been major purchasers of fire appliances since 1920.

The Ford Motor Company was probably the first manufacturer to supply specialised fire trucks to the Armed Forces, supplying the Royal Air Force with a model called 'War Office Type One', otherwise known to servicemen as a `Woti'. Three versions of this appliance are believed to have been built. A 1942 version is displayed in the Spitfire and Hurricane Museum at Manston Airport, Kent.

During the Second World War large numbers of airfield crash tenders were built using the Alvis, Bedford, Crossley, Ford and Thornycroft chassis. Today just two models serve the Royal Air Force – Mark 9 and Mark 11 – the former built by Angloco on a Thornycroft chassis, the latter with a Scammell chassis and bodywork by Gloster Saro.

Up to the early 1990s all three Armed Services had had their own fire commands and training establishments. However, on the 1st April 1991 these were brought together under the title of The Defence Fire Service, with the total requirement of specialised fire appliances running well into three figures.

Left

1942 FORDSON WOT 1 FOAM TENDER. *This 'War Office Type 1' foam tender was supplied new to the Royal Air Force in 1942. Hundreds of these appliances were built during the Second World War by the Ford Motor Company, with three variations of the design. They were designed for a crew of two and featured a Ford V8 side-valve 30-bhp engine, and a crash gearbox. In 1946, the Ministry of Civil Aviation purchased this appliance and placed it at Lydd Airfield, Kent. In 1977 it was given to the R.A.F. Fire Service School at Catterick, whose staff completely renovated it. Now on display at Manston Airfield, Kent, it is believed that only three of these appliances exist today*

1955 FORDSON THAMES 500E FIREFLY/WADHAM PUMP FOAM TENDER. *This appliance was one of 12 ordered by the Ministry of Defence for use in Royal Ordinance Factories; the one seen here is stationed at R.A.F. Barnbow, near Leeds. It is equipped with a Godiva 500-gpm pump, a 150-gallon water tank, a 150-gallon foam tank and a 30' Ajax ladder. Put out of service in 1984, it was bought for preservation and restoration in 1987. Only four of these appliances are known to have been preserved*

1957 COMMER Q4 / MANN EGERTON TRANSPORTABLE WATER UNIT.
This appliance does not of course strictly speaking constitute a military vehicle: but its raison d'être is certainly the threat of war. One of the Green Goddess family of appliances which formed the Home Office Mobile Fire Columns established in the early 1950s to combat a possible nuclear war, this water unit was staffed by members of the reformed Auxiliary Fire Service. Known as a 'Bikini Unit,' the appliance was built on a Commer Q4 4 x 4 wheeled chassis and equipped with nine portable 1,000-gpm/100-psi pumps and three inflatable rafts.

When afloat each raft carried three pumps which could then pump water ashore (the water being sucked through the bottom of the rafts) into temporary mains or portable water dam units. A form of jet propulsion was used to move and steer the rafts, with a branch pipe connected to one of the pumps. The appliance seen is believed to be the only one of nearly 300 built that is still fully equipped

1970 LAND ROVER / HCB ANGUS TAC R 1. *A total of 83 truck Airfield Crash Rescue Mk. One Appliances were built between 1968-70 by HCB Angus at Southampton. Issued to both The Royal Air Force and The Royal Navy, they were used primarily for first-line rescue duties on home and overseas military airfields. The Mk. One has since been superseded by the Mk 2A, which is based on a double rear axled Range Rover. (The Army Fire Service were equipped with their own design, constructed on a Land Rover forward-control 1 ton chassis.) The vehicle shown here (71 RN 52) served at The Royal Naval Air Station at Portland, Dorset, in the early 70s. It is fully equipped and preserved by the author*

Above

TRUCK FIRE-FIGHTING FOAM MK 9 CIRCA 1975. *This appliance was primarily designed for use as a foam crash truck, but it could also be used as a water tender, or a domestic truck. Built on a Thorneycraft Nubian Major 10 ton 6 x 6 chassis, with a wheelbase measuring 15'(4.572m) x 8' 2" (2.5m), it was powered by a Cummins V903 diesel 306 gross hp engine. Equipment included a 1,250-gallon water tank, a single-stage centrifugal MK 14 water pump (driven by power take-off from the chassis), a 130-gallon Foam Liquid Tank, a Trinity type 2500 monitor with output of 590 gpm at 180 lbs/sq in, with a minimum throw of 180'. This particular appliance was photographed at R.A.F. Gibraltar, 1983*

Right

1985 REYNOLDS BOUGHTON/MOUNTAIN RANGE MULTI-PURPOSE APPLIANCE. *First supplied to the Army Fire Service at Army Air Corps, Netheravon, on 8th February 1986, this appliance is one of only three built. Known as an R.B. 44 multi-purpose vehicle, it is fitted with a Godiva 500-gpm fire pump, a short extension ladder and a small foam monitor on the roof, front winch and telescopic lighting*

Civil Aviation

Today, international airports each have five or six major foam crash tenders, plus ancillary first-line appliances. In the age of jumbo jets carrying in excess of 400 passengers, time in reaching a crashed aircraft is measured in seconds rather than minutes.

Airport appliances are specially designed with power units which allow far greater mobility than with conventional fire appliances, and can easily handle excessive speeds along flat airport runways and over rough terrain.

During the last 20 years the number of specialist manufacturers of airport appliances has diminished. Today, Carmichael, Reynolds Boughton and Chubb Fire are world leaders in the design of these specialist vehicles, and many of their appliances can be seen at international airports in both Great Britain and around the world.

1970 THORNYCROFT NUBIAN MAJOR 6 X 6 AIRFIELD APPLIANCE.
With Carmichael bodywork built on a 6 x 6 Thornycroft Nubian chassis, this appliance is fitted with foam equipment for fighting aircraft fires. Originally stationed at Bristol Airport, it is now in preservation

Above

1971 REYNOLDS BOUGHTON CHUBB 'PATHFINDER'.
The Pathfinder is an excellent cross-country vehicle, with a top speed of approximately 60 mph. Built on a Reynolds Boughton 'Griffin' 6-wheel drive high performance chassis, it measures 37' 4" long, 10' wide and 13'7" high, and weighs 37 tons fully loaded. The water tank capacity is 3,000 gallons, plus an integral 360-gallon foam liquid tank, manufactured in fibre glass. It is equipped with a Coventry Climax 1700-1900 gpm single-stage centrifugal main water pump (later models fitted with Godiva UFP Mk 2021) and a 200-230 psi Chubb dual output monitor, which rotates through 330° and elevates 45°. In addition to one hose-reel, there are two foam sidelines and one water sideline. The vehicle carries a crew of four (although the advanced design permits a one-man operation if required). Designed for Pyrene (now Chubb Fire Vehicles) as an Airfield Crash Truck, it was first manufactured in 1971. The Pathfinder was awarded the Design Council's Engineering Award in 1974. A total of 92 had been supplied world-wide up to the end of 1982

Above right

1975 CHUBB 'SPEARHEAD' RAPID INTERVENTION VEHICLE. *Designed as a high-speed airport rescue vehicle capable of traversing rough or soft ground, 15 of these*

'Spearheads' have been supplied to the British Airports Authority. Built on a Reynolds Boughton 'Pegasus' 4 x 4 chassis, the vehicle is powered by a rear-mounted Chrysler H440 4V petrol engine and fitted with an Allison MT 640 4-speed automatic gearbox. Equipment includes a Godiva UMP Mk. 5OA 2-stage aluminium fire pump; a 200-gallon (903-litre) fibreglass water tank; a 15-gallon (68-litre) light water concentrate tank; two hand lines and a high-pressure first-aid hose reel. The 7½ ton vehicle carries a crew of four and is capable of a top speed of 65 mph

Right

1979 GLOSTER-SARO 'JAVELIN' FOAM TENDER. *The Javelin was designed in conjunction with the British Airports Authority in 1979 as a high-mobility airfield crash tender and fire-fighting appliance. Its aluminium modular-unit body was built on a 6-wheel centre-drive Boughton 'Taurus' chassis, and fitted with a 5-speed Boughton/Borg Warner automatic gearbox and a 13.94 litre turbocharged V12 Detroit diesel engine, producing 55.3 bhp at 2,100 rpm. The Gloster-Saro foam monitor is capable of discharging 10,000 gallons of aerated foam and water per minute. Later models were fitted with a Sky King hydraulic platform rising to a height of 34'5". Other equipment includes a Godiva Mk 20 single-stage centrifugal pump, various air tools and elevated floodlighting. Designed for a crew of four*

Above

1984 CHUBB 4 X 4 PROTECTOR AIRPORT CRASH TRUCK. *First supplied to Teesside International Airport in 1984., this fast-response 4 x 4 Airport Crash Truck is capable of achieving 0-50 mph in 28 seconds and producing 4,000 litres of foam per minute*

Left

1987 CARMICHAEL AIRFIELD CRASH TENDER. *This vehicle is a purpose-designed water-foam-BCF Crash Tender comprising the most sophisticated equipment available. Featuring a Perkins TV8 540 diesel engine, with a power output of 215 bhp, the Crash Tender is fitted with an Allison MT 643 automatic transmission. Special equipment includes a Godiva UFPX 660 main fire pump, delivering 1,000 gpm; and the Carmichael 4000 monitor with a water output of 400 gpm, capable of throwing foam up to 140-150 feet depending on wind conditions. It has a 600-gallon water tank, a 72-gallon foam tank and two 50kg BCF units. Built in 1987 by Carmichael Fire and Bulk Ltd, Worcester, it was supplied to the then new London City Airport*

Ladders

Fire-fighters' ladders were of course originally designed for rescuing people, but today's hydraulic ladders and aerial platforms – which can be raised to over 100 feet in a matter of seconds – are used both as a rescue device and as a static water tower. The ladder's variable booms can be manoeuvred high over the seat of a fire, as water is pumped to the head of the ladder and discharged through a monitor.

The short extension ladder – seven or eight rungs high – was first introduced on the fire-fighter's hand cart. Later, longer ladders were used with their own horse-drawn wagons. Around 1836, 55-foot ladders were produced, with wooden-wheeled axles and a canvas escape chute attached. These chutes were known as 'Wivel Fly Ladder Escapes', and were the forerunner of the wheeled escape, which was to remain in service for another 130 years.

By the turn of the century the wheeled escape had become an integral part of the motorized appliance. The first wheeled escapes were made entirely of wood, but later Merryweather introduced an all-metal assembly with metal wheels and thin solid rubber tyres.

In 1908 Merryweather introduced their petrol engine turntable ladder. (Earlier types of ladders had been elevated by compressed carbon dioxide gas; however, when the gas ran out, the ladder had to be manually wound down.) This new ladder was operated by only two levers, the chassis being fitted with four stabilising jacks. The first model was exported, but was later supplied to British fire brigades as well.

In 1933 Merryweather exported their first all-metal ladder to Hong Kong. A few years later, British manufacturers were fitting the much-

1949 DENNIS F7 PUMP ESCAPE. *The F7 was most notable for its acceleration time of 0–60 mph in just 45 seconds! Powered by a Rolls Royce B.80 Mk X Petrol engine, the appliance was fitted with a 4-speed manual gearbox, a Dennis No. 3 pump with two power take-offs, and a first-aid pump. This design was the forerunner of a range of familiar 'F' series specialist appliances to be delivered to British Brigades. London would take the first in 1949. (The F7's 162" wheelbase was soon superseded by one 12" shorter, the F12.) This particular F7, seen here carrying a 50' Bailey wheeled escape ladder, went into service at Caversham Road Fire Station, Reading, in 1950, where it served for 23 years. It is now owned by the Berkshire Veteran Fire Engine Society*

acclaimed German manufactured Metz and Magirus ladders; however, the supply was cut off at the start of the Second World War. Luckily, Merryweather were equipped to manufacture the larger, four-sectioned, turntable ladders, which were fixed to the Leyland and Dennis chassis, with many being supplied in the wartime grey livery to the newly formed National Fire Service (NFS).

During the Second World War, Merryweather continued producing a limited number of 100-foot turntable ladders, but priority was given to the 60-foot manually-operated ladder which was fitted to the Austin A4 chassis. These appliances were considered the backbone of the NFS.

Prior to the Allied invasion of Europe in 1944, under cover of great

Above
TYPICAL LOCKER LAYOUT OF 'F' SERIES APPLIANCE

Right
DENNIS 'F8' WATER TENDER LADDER. *Over 150 of these models were built at Guildford between 1948 and 1957, equipped with Rolls Royce B 60 petrol engines which had a power output 175 bhp. The F8 featured a manual 5-speed gearbox, a Dennis No. 2 water pump delivering 500 gpm at 100 psi, a single 180' hose reel, an Ajax 35' ladder and a 200-gallon first-aid water tank. Later models were fitted with a 10' x 6' 6" wheelbase and carried 250 gallons. After being stationed first at East Preston in 1956, and later at Shoreham by Sea, this Dennis appliance was purchased by West Sussex Fire Brigade Charities in 1976*

1950 AEC REGENT Mk. 3 DUAL-PURPOSE PUMP ESCAPE. *Built on a double-decker bus chassis, this dual-purpose appliance was fitted with a Merryweather 50' metal wheeled escape, a Merryweather 1000-gpm pump and a separate first-aid reel pump, and was powered by a 9.6 litre, six-cylinder AEC type A 218 engine. No. 32 of a total of 115 built, this was the first high-speed diesel engined appliance in the North of England. Ordered by South Shields Fire Brigade in 1950, it was completed by Park Royal Vehicles Ltd and exhibited at the 1951 Earls Court Commercial Vehicle Show. It remained 'on the run' until 1968, and was then used for driver training. In 1974 the vehicle was purchased by Mr Paul Pearson, who has restored it to the '50s era with over 250 items of tested equipment*

1951 DENNIS F12 PUMP ESCAPE. *This was mechanically identical to the Dennis F7 but with a 12" shorter wheelbase. It was one of the largest coach built appliances produced at Guildford in 1951, and was apparently quite popular, since production continued for nine years. Featuring a straight eight-cylinder Rolls-Royce 5,996cc petrol engine, developing 160 bhp (36.9 RAC), this model was fitted with a 4-speed manual gearbox and two power take-offs for a Dennis No.3 pump (1,000 gallons/4,670 litres per minute), and a first-aid pump (50 gallon/235 litre capacity). It is seen here carrying a Merryweather 50' wheeled escape ladder. Having been stationed at Salisbury until 1978, it is now preserved by Wiltshire Fire Brigade*

secrecy, a number of Merryweather's standard four-sectioned fire ladders were mounted onto naval landing craft – not for fire-fighting purposes, but rather to be used for lifting assault troops up the cliff faces of the Normandy beaches .

In the early 1950s, modern hydraulic technology was applied to turntable ladders. Today some fire officers still choose these appliances, whilst many others prefer the longer-reach aerial platforms, the majority of which are supplied by Simon Dudley Limited of the West Midlands. During the last decade great advances have been made in fire appliance technology. The Bronto Skylift is the undisputed leader in aerial platforms, most fitted to the Volvo or Scania purpose-designed chassis.

The Bronto Skylift - and the transformation in 'aerial' technology in general - is just one aspect of technical improvement and specialization. Today's modern fire brigades incorporate a range of emergency and specially designed vehicles because of the wider responsibilities undertaken by fire-fighters, ranging from rescue work at major motorway incidents and gas explosions, to industrial and aircraft disasters. Some of these vehicles are illustrated in the following chapter.

The cost of the sophisticated equipment carried by today's emergency rescue trucks is equivalent to the cost of the truck itself. Powerful hydraulic cutting equipment, a heavy lifting crane, airbags, generators, extensive high-powered lighting, hydraulic jacks and many trays of specialist hand tools and accessories are modern necessities. In addition, water tankers are required to be carried in areas where local supplies may be insufficient.

Mobile control centres, which supply facilities for serving hot drinks and meals, are sometimes used in areas where over six fire appliances are in action. (Records show that London Fire Brigade had a mobile canteen in use in 1908.)

1987 M.A.N./METZ HYDRAULIC TURNTABLE LADDER. *Originally supplied to East Sussex Fire Brigade in September 1987, this appliance is now stationed at Eastbourne on the south coast. Bodywork is by Angloco; M.A.N. chassis having a tilt cab. The 100-foot all-metal four-sectioned Metz ladder is fiited with a working cage and is able to hold a stretcher case. The crew of three can use the boom as a crane jib to lift 3000 kg at 60°*

Above
1939 AUSTIN K4 ESCAPE Ladder. *The manually operated Austin K4 escape ladder is in wartime livery, here standing next to the latest 1993 Volvo/Bronto Skylift 28-2TI hydraulic platform; (see pp 126-7)*

Above right
1956 QX DUAL-PURPOSE TYPE 'B' WATER TENDER ESCAPE. *Built by Alfred Miles of Cheltenham for Somerset Fire Brigade, this was stationed at Minehead until 1969 and then served as a driver training vehicle until 1975. Purchased by its present owner in 1976, the vehicle has been completely re-equipped to the original specification. It has a six-cylinder petrol engine, a 4-speed gearbox, a Dennis no 2 pump, a 400-gallon water tank, two hose reels, a 35" Ajax ladder, and, when fitted, a 50' wheeled escape. It was designed to carry a crew of six*

Right
1956 DENNIS F 101 DUAL-PURPOSE PUMP ESCAPE. *Powered by a six-cylinder Rolls-Royce C 6 diesel engine, the F 101 was Dennis' first diesel fire appliance and is the only working survivor of 38 built for the London Fire Brigade. The model was fitted with a 4-speed reverse gate crash gearbox and equipped with a Dennis No. 3 1,000-gpm pump, a 100-gallon water tank, a 50' wooden Bayley wheeled escape, two hook ladders and two scaling ladders, three sets of Proto breathing apparatus, and two low-pressure hose reels. The appliance served the Fire Brigade until 1979 and was then sold for preservation. It has since been extensively reconditioned and restored to 1956 livery by its present owners*

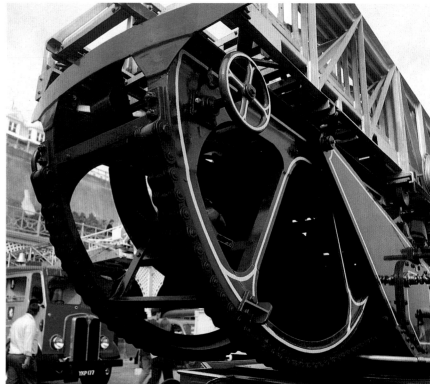

Above left

1958 BEDFORD HCB 'S' SERIES PUMP ESCAPE. *Built at HCB Southampton, this appliance was supplied in 1958 to the Northern Area Fire Brigade. It was stationed at Lerwick on the Shetland Isles, also served on Stornaway and Golspie, and retired in 1983 having travelled only 3,000 miles. Powered by a Bedford six-cylinder petrol engine and gearbox, the appliance is fitted with a 500-gpm Coventry Climax pump, a 250-gallon tank and a 50' Bayley wheeled escape*

Left

POWER AND STRENGTH
The fulcrum frame of a 104ft Metz turntable ladder.

Far left

1988 HESTAIR DENNIS / CARMICHAEL / 100FT MARIGUS TURNTABLE LADDER

Above

1937 LEYLAND OPEN TURNTABLE LADDER PUMP.
Supplied new in 1937 to Shaftesbury Avenue (Soho) Fire Station, London, this mighty vehicle is 33' long, 7'6" wide, 10'6" high, and weighs 9½ tons. Its 4-section steel Metz ladder extends to 101'. Powered by an 8.8 litre Leyland OHC six-cylinder petrol engine, it was designed to carry a crew of two (three during wartime). and was fitted with a 500-gpm Rees-Roturbo pump. Other equipment includes two hook ladders, a monitor fitted to the head of the ladder, and trolley bus short-circuiting equipment. This appliance has appeared three times in the annual Royal Tournament at Earls Court, London. After 26 years service in Soho, it was purchased in 1981 by Mr Mike Hebard who has restored it to original L.C.C. livery and full working order

1959 BEDFORD 'S' TYPE DUAL-PURPOSE WHEELED ESCAPE. *This appliance was specially built by A.E. Smith of Kettering to serve the villages of Brackley and Oundle, Northamptonshire, both of which have narrow winding streets and three-storied public schools. Along with a Bedford six-cylinder petrol engine, this model was fitted with a 4-speed gearbox, a Hayward Tyler 500-gpm pump, a 500-gallon water tank, two first-aid hose reels, a 50' wheeled escape and assorted ladders. It was designed to carry a crew of six. In 1979 the appliance was sold and has since been preserved*

1938 LEYLAND TURNTABLE LADDER. *Fitted with an 8.8 Leyland engine and a 4-speed Leyland wide ratio gearbox, this appliance features a Merryweather mechanical turntable ladder which extends to 100 feet. Weighing in at 9 tons 14 cwts, it was supplied to Bridewell Fire Station, Bristol in 1938. In 1939 it moved to the City of Salisbury, and was painted NFS grey in 1941. After the war it was repainted red, and remained in service at Salisbury until 1968. It was then sold for scrap and is now privately owned*

1954 LEYLAND BEAVER TURNTABLE LADDER PUMP.

This massive appliance is 33½' long, weighs 11 tons and is powered by a Leyland 0600 9.8 diesel engine, fitted with a 5-speed crash gearbox. Designed for a crew of three, it is equipped with a 100' 4-section steel mechanical German-built Magirus ladder and a 1939 Coventry Climax FF pump (capable of 300-500 gpm), similar to that of the wartime trailer pump. One of only two built for UK use, it was supplied in March 1954 to West Riding County Fire Brigade at cost of just over £8,500. Bodywork by John Morris & Son, Salford, was set on a Leyland Beaver lorry chassis. The appliance first served Batley until 1964, then Keighley until 1970, during which time it caught fire resulting in a major overhaul. Finally, it served at Brampton until 1973. It was sold in 1974 to Yorkshire Fire Museum, and then sold for continued preservation in 1984. This photograph was taken in 1989 outside Netley Royal Victoria Military Hospital Museum, Southampton. This appliance was recently purchased by David Plant, whose father Alfred Plant was part owner and a director of John Morris and Company at the time of the vehicle's production

1966 MERRYWEATHER HYDRAULIC TURNTABLE LADDER.

The Merryweather 100' (30m) all-British hydraulically-operated turntable ladder comprises a steel ladder in four sections, mounted on an AEC diesel-driven chassis. Special features include a self-contained diesel-driven pumping set, which incorporates a Merryweather single-stage pump capable of 300-600 gpm; and a desk-type console mounted on the turntable which revolves with the ladder. When fully closed the ladder may be used as a crane jib, lifting a maximum load of 2,000 lbs at 60°. The cab carried a crew of six; the wheelbase of the 9½ ton vehicle measures 15' x 7'6"

Above

EPL FIRECRACKER 235 FIRE-FIGHTING AND RESCUE PLATFORM. *The boom on this aerial platform, specifically designed for fire-fighting, is mounted to give 360° continuous rotation. The cage, which has a fold-up rescue platform, can reach 78 feet (23.3 metres) and a maximum outreach of 37'6" (11.5 metres). Under the cage is a spray nozzle which provides a vertical water curtain. This model was individually produced by EPL International in the late 1970s and early '80s*

1979 HESTAIR DENNIS DELTA II HYDRAULIC PLATFORM.
The Simon SS 220L snorkel unit is mounted on a Hestair Dennis purpose-built chassis, and powered by a Perkins V8 640 diesel engine, developing 215 bhp at 2,600 rpm through an Allison MT 640 automatic gearbox. The hydraulically-operated boom gives a working height of 22 metres and an outward reach of 12.45 metres

1982 BEDFORD CARMICHAEL TURNTABLE LADDER. *Designed to carry a crew of three, this modern appliance measures 28'6" long, 8'1" wide and 10'7" high. Built in 1982 by Carmichael, and supplied to Oxfordshire Fire Service, it is currently 'on the run' from Rewley Road, Oxford. It is powered by a Bedford 8.2-litre 500 Turbo diesel engine, fitted with an Allison 643 Auto gearbox and equipped with a Godiva GT 2,200-gpm pump and a Magirus DL 30 (30-metre) all-steel ladder*

Left

WOODEN WHEELED ESCAPE — CIRCA 1960. *This is the site of one of the last major fires in London where wheeled escapes were used. At Chappells music and piano shop on New Bond Street, sheet music fuelled a fierce fire, resulting in the death of two staff. Being the first photographer to reach Fleet Street with exposed film – just as news of the fire was being received on the teleprinters – it was suggested in jest that perhaps I had started the fire! The photographs made front page of* The Daily Telegraph *the next morning*

Above

1962 BEDFORD/HCB WATER TENDER TYPE 'B'. *Powered by a Bedford 300 petrol engine, this Type 'B' model was fitted with a Hayward Tyler 500-gpm pump, a 400-gallon water tank, two hose reels and a 35' Bailey ladder. It was first supplied to Lancashire County Council at a cost of £32,800 (excluding equipment), and was stationed at Coniston in the Lake District. In 1974 the Coniston area was absorbed into the Cumbria Fire Service, and the appliance was sold at public auction in 1982*

Modern Machines

The post-war years have been outstanding for British fire appliance production, with significant changes and improvements in design, safety, bodywork construction and motive power seen from 1946 to the present day.

In the late 1970s the classic outline of the '30s and '50s appliances was to end with the introduction of the Crew Safety Vehicle (CSV) pioneered by HCB Angus. The basis for this design was a rigid steel tubular framework to protect the crew in the event of a roll-over accident; however functional, the vehicle resembled a box on wheels. 'Arrive alive and alert in the HCB Angus CSV' was the advertising slogan.

Dennis of Guildford, a well-established pre-war company (founded in 1895 by cycle manufacturing brothers, John and Raymond) launched into the post-war years by introducing the 'F' series of appliances. After the F1 and F2, the F7 water tender built in 1949 was much admired, and that design was to be followed by a 12" shorter wheel base, the F12, which was powered by a Rolls-Royce 880 Mk.10 petrol engine. During the next 10 years or so, many F models followed, with the slightly narrower-bodied F8 becoming widely accepted throughout the UK fire brigades. The F8 later became much sought-after by preservationists, many of whom regard the F8 as a classic vehicle.

In the early '60s, Dennis introduced the Delta chassis, which was fitted with the 'Simon Snorkel', a very manoeuvrable hydraulically-operated platform, serving as a replacement for the traditional turntable ladder. Today, Simon-Dudley Engineering Ltd (W Midlands) are one of

1978 DENNIS 'RS' FIRE APPLIANCE. *The Hestair Dennis 'RS' design, introduced in 1978, is based on front-line action experience gained from the world-wide use of 'R' series appliances. Powered by either a Perkins 640 or 540 V8 form diesel engine, it is fitted with a choice of a fully automatic Allison MT643 gearbox or Turner T5 400 5-speed manual synchromesh gearbox (V8 540 engine only). Most vehicles were fitted with the Girling Skidcheck system as a standard option. Equipment includes a No. 2 Dennis 500-gpm main pump, a 45' Angus 464 alloy ladder and other alloy ladders, and a 400-gallon water tank. Five B.A. sets fit behind the crew seats. The driving and crew cab, common to all models, is of an all-steel construction designed to meet modern impact and crew safety regulations. With a wheelbase of 12'6", the fire appliance is 23'9" long, 7'6" wide, 10'1" high and has a gross weight of 11 tons*

the leading producers of this type of appliance, both for fire brigades and for industry.

Of the many other appliance manufacturers in business at the time, some continued successfully and others dropped out of what was to become a very competitive industry. Merryweather, who moved from their Greenwich Works in London to Ebbw Vale in Wales and later to Plymouth, Devon, were soon to disappear after 293 years of pump production. Leyland gave up after a short-lived post-war revival, although the new Leyland chassis is now being used by several UK fire brigades.

Above
1963 BEDFORD TK WATER TENDER. *The Bedford TK Water Tender features a 400-gallon water tank, a 600-gpm fire pump and a 35' Bayley ladder. Originally supplied to Dorset Fire Brigade, it was sold in 1979 to Flight Refuelling Ltd of Wimborne, Dorset, for use by their Works Fire Team*

Right
1967 DENNIS 'F44' WATER TENDER LADDER.
This is one of the famous Dennis 'F' series built at Guildford. (Although over 1,400 'F' series appliances were produced, only about 30 were F44s, production of these being between 1967-71.) Powered by a B81 Rolls-Royce straight eight-cylinder petrol engine, developing 235 bhp at 4,000 rpm, the appliance was fitted with an Allison Automatic transmission and had a 12'6" x 7' wheel base. Equipment included twin 180' hose reels, a 400-gallon first-aid water tank and a 45' ladder

Around 1953, with the threat of nuclear war developing, Bedford and Commer benefitted from orders placed for large numbers of Green Goddess pump appliances, pipe carriers, 'bikini units', general-purpose transport, and control and kitchen vans (see pp 66-67). This equipment was to form countrywide back-up units staffed by the reformed AFS (which, after weekly musters and many nationwide training exercises, were again stood down). It has been estimated that around 5,000 vehicles were produced in response to the Cold War.

Eventually, all of these pieces were auctioned off, with the exception

Above
1941 AUSTIN H.P.U. REBUILT. *Originally supplied new as a wartime heavy pump unit to Coventry Fire Brigade, this appliance was later rebuilt with new bodywork and is a typical example of how wartime appliances were revitalised after the war. After being sold to car manufacturers Jaguar/Daimler for their Works Brigade, it passed to The Museum of British Transport, Coventry, for preservation*

Left
1967 COMMER OX/CARMICHAEL WATER TENDER. *Supplied to East Riding Fire Brigade in 1967, this appliance served at Pocklington and Bridlington, was transferred to Humberside Fire Brigade in 1974, and served at Beverly until 1984. Powered by a Commer 4.75 litre engine, developing 109 bhp, and fitted with a manual 4-speed gearbox, it carries a Merryweather 350-gpm pump, a 400-gallon water tank, two hose reels, a 35' Ajax ladder, a portable pump and foam equipment*

Above

1970 MK. TWO DENNIS / J.C. BENNETT 'SCOOSHER'. *When introduced by Glasgow Fire Service in 1969, the Mk. One was described as 'one of the most exciting developments in recent years' and 'the appliance of the 70s'. It was built on a Dennis 46A type chassis, powered by a Rolls-Royce eight-cylinder 235-bhp petrol engine, and fitted with a Dennis No. 3 1,000-gpm pump and a 300-gallon fibre glass tank, providing an initial supply to the pump. Coachwork was by J.C. Bennet & Co (coachbuilders) of Glasgow. The concept behind the 'Scoosher' was to have a fast, adaptable appliance, equipped with a versatile turntable boom, fitted with a monitor which had an AFA Infrastat infra red detector fitted alongside the monitor itself, thus using the detector to seek out the source of heat in openings above ground level or in smoke logged areas, in turn directing the monitor's one-inch jet or fog spray. Wide horizontal and vertical sweeps were possible; the maximum height obtained from the two-sectioned boom was 45' (13.7 metres). Sadly, within 10 years all 14 appliances built were withdrawn from service. The appliance seen here is now the only one left, safely preserved in private ownership. By the way, the word 'Scoosher' is an old Scottish name for a water pistol*

Right

1969 DENNIS 'D' SERIES WATER TENDER LADDER. *Boasting a Rolls-Royce B 81 SV engine, this vehicle has a 5-speed close ratio gearbox and is capable of developing 235 bhp at 4,000 rpm. Built to carry a crew of six, the 23' x 7'6" x 10' vehicle is equipped with a Dennis 1,000-gpm pump, a 400-gallon water tank, a 464 Ajax ladder and several short extension ladders. This particular vehicle was built in 1969 and supplied to Suffolk Fire Service, where it was converted in the Brigade workshops, the rear end enlarged to store additional accident gear, a generator, cutting equipment and chemical suits. It was sold to Felixstowe Docks in 1983*

1971 LAND ROVER/BRANBRIDGE LOW-PROFILE WATER TENDER. *Here we see an example of the versatile 4-wheel-drive Land Rover. This vehicle was specially designed for Cheshire Fire Brigade for use in the Runcorn Shopping precinct. Many similar appliances were built with maximum height restrictions being of prime importance, in order to enable easy access to pedestrian areas and multi-storey car parks. Three models were available for first-strike use, all chassis being modified with heavy-duty rear suspension. An interesting feature of this appliance is the front bumper, which was re-designed to incorporate an extra water tank, acting at the same time as front-end ballast*

1971 FORD/PYRENE FOAM TENDER. *Built by Pyrene (now Chubb Fire Vehicles) for London Fire Brigade in 1971, this vehicle features a Ford Perkins 511 V8 diesel engine, developing 170 bhp at 2,800 rpm, and a Turner TSC 4008 gearbox. Constructed of welded steel frames, panelled in light alloy, and mounted on a Ford D 1000, 159" wheelbase with a tilt cab, it was designed to carry a crew of two. Equipment includes a Dennis No. 3 two-stage turbine fire pump; an 800-gallon foam compound tank; an Albany A.P. 11.5 foam compound pump; a Pyrene jetmaster monitor; and four F.B. 10 foam-making branch pipes, designed to produce a potential foam output of 5,700 gpm*

of the Green Goddess pumps, which were placed in storage for future emergencies. The pumps did indeed come into service – they were used during the Firemen's National Strike in 1977, and more recently they pumped torrential flood waters from the streets and cellars of the Sussex cathedral city of Chichester in January, 1994.

Other well-known post-war manufacturers included John Morris; Thornycraft; ERF; Mountain Range; Shelvoke and Drewry; Fulton and Wylie; and Locomoto. All have ceased production of fire appliances, and at the time of writing, the highly-respected company HCB Angus at

Left
1979 SHELVOKE SPV/CHESHIRE FIRE ENGINE PUMP ESCAPE *This appliance, supplied to the London Fire Brigade, started its operational life at Wembley, and finished at Chelsea where it was the last appliance in London to carry a wheeled escape*

Above
1972 CARMICHAEL/RANGE ROVER 'COMMANDO' RAPID INTERVENTION VEHICLE. *Introduced in 1972, this model was built on a standard Range Rover with a trailing third axle added by Carmichael and Son of Worcester. Using a Rover V8 petrol engine developing 130 bhp at 4,700 rpm, it was fully equipped for emergency rescue work with a generator, a telescopic lighting mast, a winch with a 100' long x ⅝" cable, portable cutting tools, an Albany AP8 pump and a 490-litre pre-mixed concentrate light water tank*

Above
1952 MORRIS MINOR FIRE APPLIANCE. *This one-off appliance started life as a prototype chassis for the Morris Minor Van. It was then built in its present form to become a first-line fire appliance at the Morris Cowley Motor Works, Oxford, being just narrow enough to drive between the vehicle assembly lines. It can now been seen at The Heritage Motor Centre at Gaydon, near Warwick*

Left
WHEELED FIRE EXTINGUISHER CIRCA 1950. *Many types of fire extinguishers have been manufactured by nearly as many companies. Large models, as seen here at Felixstowe Docks in the 1960s, were mounted on metal wheels and known as 'chemical engines'. Those dispensing foam were cream-coloured, and the soda acid containers were red. These were a familiar sight in military establishments*

Southampton have announced that they too will cease trading as of May, 1994 – a very sad chapter of events indeed.

Amongst the appliances to end the selection of photographs in this edition, I have chosen the Dennis water tender 'Rapier' (see page 124), much acclaimed for its superb road handling, with coachwork by John Dennis Coachbuilders Ltd, also at Guildford. (John is the grandson of John Dennis, one of the founders of the original company.) The giant pictured on pages 126-7, is the most expensive – Merseyside's Bronto Skylift, fitted on a Scottish-built Volvo chassis, weighing in at 22.5 total tons with a price tag of £400,000.($600,000). These examples of British fire appliance technology are a tribute to the expertise which has led the industry for many years and which will no doubt continue to lead.

Above

1976 ERF PUMP ESCAPE. *This ERF type 84 pump escape was based at the Barbican in London and later in Peckham and Heston. To the author, this is the classic shape, before appliances took on today's boxed look*

Right

1957 LONG WHEEL BASE LAND ROVER AND 1971 LAND ROVER WATER TENDER . *Since its introduction in 1948 the 4-wheel-drive Land Rover has been used extensively as the basic unit for many different types and designs of 'first-strike' fire appliances and has been exported as such to many parts of the world. The vehicle shown on the left was supplied to West Sussex Fire Brigade in 1957. It is equipped with a Hathaway 88/80 pump which is fed from an 80-gallon water tank. The appliance on the right was unique in the West Sussex Fire Brigade, being a 110 forward-control Land Rover with bodywork by Carmichael and Son of Worcester. Powered by a six-cylinder petrol engine with 4-wheel-drive, the model carries 455 litres of water and is fitted with a 1,365-lpm Coventry Climax pump. The appliances seen here are in the yellow livery recommended for fire appliances in 1971*

Above

1993 SCOT TRACK ALL-TERRAIN VEHICLE. *This rugged vehicle was purchased by North Yorkshire Fire and Rescue Service for carrying the Moors Pack, consisting of two back-pack tanks to control small outbreaks on Moorland. It is fitted with an inflatable shelter for use when working in remote areas*

Left

1976 HCB-ANGUS CSV TYPE 'B' WATER TENDER. *The result of practical full-scale testing, this vehicle was designed to be the ultimate in basic fire-fighting appliances, incorporating a new standard of crew safety. Built on a Bedford KGS forward-control 4 x 2 chassis, the all-metal bodywork incorporates a driving/crew safety cab seating six. The entire vehicle measures 23' in length, 7.5' in width and 10' in height. It is powered by a Bedford six-cylinder, 8.2-litre, 160-bhp diesel engine and fitted with a 5-speed manual gearbox. The rear-mounted Godiva UMP Mk. 50A fire pump is driven by the engine, delivering 500 gpm at 100 lbf/sq in. The pump is fitted with a 4" suction coupling and either two, three or four 2½" delivery valves. Other equipment includes a 400-gallon (1,820-litre) water tank; two HCB-Angus 180'(55m) high-pressure first-aid hose (¾") reels; an Angus '464' 45' (13.7m) three-section light alloy extension ladder; a 16' (4.9m) roof ladder; and a 21' (6.4m) three-section ladder. There are three lockers on each side, fitted with slat-type rolled shutters, and the crew cabin features HCB-Angus 'sitdown/strap on' breathing apparatus modules*

Above

1985 DENNIS 'SS' WATER TENDER. *This is a sister to the Dennis 'RS' pictured on page 98, the 'SS' being offered with the same power units and equipment, but with the addition of a tilt cab. A total of 2,300 Dennis RS and SS appliances have been built since 1978. At the time of writing, production of these two models is apparently to cease at the end of 1994*

Left

1976 DENNIS 'R' SERIES WATER TENDER LADDER. *The successor to the 'D' series, just over 200 of these were built at Guildford between 1976-79. Powered by a Perkins V8 540 diesel engine developing 178 bhp, the 'R' series appliance is fitted with a Turner 5-speed manual gearbox. Equipment includes a Dennis No. 2 water pump (capable of 2,140 lpm or 10.5kg/sq cm), 180' Auxiliary twin hose reels and a 1,850-litre first-aid water tank. The wheel base measures 11'9¾" x 7'6". The appliance seen here was supplied to West Sussex Fire Brigade and stationed at Chichester*

Above

1988 VOLVO FL6 / MOUNTAIN RANGE WATER TENDER. *This sturdy appliance is built on a Volvo FL6 chassis and fitted with a turbocharged Volvo diesel TD61F engine, which is claimed to provide constant power for water pumping and a power source whilst remaining stationary for many hours. The chassis is manufactured in Belgium. The powerplant has become widely used for fire appliance production since its introduction to the British market. In addition to ladders and hoses, these vehicles carry a wide range of emergency equipment, including a Honda 3,000kw portable generator, a Makit saw and disc cutter, hydraulic spreaders, rams and jaw cutters, and a portable decontamination shower for personnel. It is also fitted with the built-in 'S' triple foam system, with a 60-litre ready-mixed foam tank. Later models are fitted with a front-mounted power winch. This particular vehicle was supplied to Buckingham Fire and Rescue Service, being first stationed at Boughton Fire Station, Milton Keynes*

Right

1984 DODGE SAXON 'COMMANDO' G 13 W.T.L. *Designed for a crew of six, this appliance is powered by a V8 engine, delivering 180 bhp at 2,600 rpm. It has a 6-speed O/d manual gearbox, and is equipped with a Godiva UMPX 2,250-1pm pump, two hose reels, an Angus 464 ladder, a Clark telescopic lighting mast, and two 500-watt Quartz Halogen lamps. Supplied new to Buckingham Fire Brigade, it is stationed at Aylesbury*

Above left

1978 FIRE BRIGADE CONTROL UNIT. *This was built for East Sussex Fire Brigade by Anglo Coach Builders in 1978. When shown new at County Hall, it is reported that a county councillor remarked that it was the most expensive worker's tea bar he had ever seen. However, 16 years later it is still on operational stand-by. Powered by a 6-cylinder 500 cu in diesel engine, this 8 ton, 17 cwt (laden weight) vehicle is 30' long, 8'2" wide and 11'7" high. Built on a Bedford chassis, the vehicle contains three compartments: a communications area with two consoles, information boards and map drawers; a conference area; and a canteen area fitted for cooking and serving refreshments. Both 24 volt and 12 volt electric power is supplied by batteries charged by a built-in generator, which also gives a 240 volt supply*

Left

1980 EMERGENCY RESCUE TENDER. *This rescue tender was built as a prototype by Pilcher Greene of Burgess Hill in 1980, on a 320/40 Jeep chassis, to the specification of East Sussex Fire Brigade. Powered by a 3060 cc American Motors engine, developing 175 bhp, it was fitted with a 3-speed automatic gearbox. Equipment for the crew of three included an 8,000-lb pull Bulldog winch, a 2.5-KW 110-volt Dale portable generator, 2-KW Dale floodlights mounted on a mast which expanded to 10' high; and a 20-cubic-feet-per-minute Hydrovane air compressor with an air hose reel fitted with four outlets for emergency lifting and cutting equipment*

Above

JAGUAR/CHUBB 'THRUST TWO'. *Regarded as the world's fastest fire appliance, this was converted by Chubb Fire Vehicles from a Jaguar XJ 12 Saloon Car. In 1981 it shadowed Richard Noble's attempt at the world land speed record at Utah, U.S.A., as an emergency vehicle. In 1982 it established an official average speed of 129.018 mph, fully laden over the measured mile at Black Rock Desert, Nevada, U.S.A. Since returning to England, it has covered major events at Silverstone motor racing circuit. The vehicle is equipped with 90 litres of pre-mixed foam, which it is able to project over an area of 110 sq m. This unique vehicle can now be seen at the Museum of British Road Transport, Coventry*

Above

1989 GMC PUMP. *Primarily designed as a first-line appliance for domestic fires, this type of vehicle has proved popular due to its size and reduced capital and running costs. A V8 6.2 GMC diesel engine runs the unit; equipment includes a 500-gpm Godiva fire pump, a 200-gallon water tank and a 10.5m Angus triple ladder. This vehicle is currently with the Buckingham Fire and Rescue Service*

Left

1991 FORD CARGO 1318/CARMICHAEL PUMP WATER CARRIER. *Delivered new to Guernsey Fire Brigade at St. Peter Port, this pump water carrier has a capacity of 4,500 litres of water and 135 litres of foam. Costing £63,883, the appliance is fitted with a 5,300-litre high-capacity fire pump*

1993 DENNIS 'RAPIER' WATER TENDER. *Designed and built by Dennis Specialist Vehicles, the Rapier chassis is described as 'the driver's machine', said to give the highest standard of road holding, manoeuvrability, acceleration and braking. Constructed with a rigid 'space frame', it is designed lower than a conventional chassis, affording optimum handling and a lower roll rate, which is achieved by the fitting of smaller wheels and tyres, giving a lower centre of gravity. A racing-car design has been incorporated to ensure maximum road holding. Powered by a choice of Cummins 'C' series diesel engines, producing up to 240 bhp, the vehicle also features a fully automatic Allison World series 5-speed gearbox. Designed to give fire crews easier access, the lower single-step entry also provides improved safety for crew. The appliance shown here was crafted by John Dennis Coachbuilders Limited, and supplied with 16 other Rapiers to Surrey Fire Brigade*

1993 VOLVO / BRONTO SKYLIFT 28 – 2 T 1. *A sophisticated combined platform and ladder aerial appliance costing £400,000, this Volvo/Bronto Skylift is built with a heavy duty box-frame chassis designed to take all stresses caused by the use of the machine. Four hydraulic outriggers form a stabilising system to ensure safe operation. It is powered by a Volvo TD 103E/ES engine with an MT 700 series gearbox. The turntable rotates through 360°, the boom reaching a maximum height of approximately 97' (29.5m) with a safe working load of 800 lbs (400 kg). The boom's movements can be controlled either from the chassis deck or from the boom's cage. The chassis and ladder weigh approximately 22.5 tons; the entire vehicle measures 32' (10m) in length, with a full working width of 18' (5.5m). Supplied to Merseyside Fire Brigade in 1993, the vehicle is crewed by the driver and officer in charge*

'FORGOTTEN'. *Found in a Sussex breakers yard, this 1942 Dennis with its Merryweather 100' turntable ladder was stationed at Brighton Fire Station during the Second World War, and traces of National Fire Service grey paint can still be seen! This photograph is dedicated to all the enthusiasts, both owners and followers, who support the preservation of redundant fire appliances; their hard work and interest ensures that many fine historic machines will be seen for a long time to come*

British Fire Engine Collections

Here is a list of interesting collections in the UK, which are well worth a visit. It is advised that you telephone to check opening times, and to make an appointment if necessary.

BANWELL FIRE STATION COLLECTION, Banwell, Somerset.
A well-preserved collection of 18th- and 19th-century manual fire pumps. Appointment necessary.

FIRE FIGHTING GALLERY, SCIENCE MUSEUM, Exhibition Road, South Kensington, London.
Scale models are also featured in this well-presented display.

GREATER MANCHESTER FIRE SERVICE MUSEUM, Maclure Road, Rochdale, Lancs.
This museum covers the history of fire-fighting in general, but particularly that of Greater Manchester, and features 16 appliances.

LONDON FIRE BRIGADE MUSEUM, 94a Southwark Bridge Road, London.
A comprehensive collection of fire-fighting equipment and memorabilia, including 25 appliances.

MUSEUM OF FIRE, Lauriston Place, Edinburgh, Scotland.
A splendid display of appliances and set pieces. Historic headquarters of the oldest UK fire brigade, formed in 1824.

SANDRINGHAM MUSEUM, Nr. Kings Lynn, Norfolk.
A small but interesting exhibition, including a preserved Royal Family Merryweather fire engine. Housed in the original fire station.

SOUTH YORKSHIRE FIRE SERVICE MUSEUM, West Bar Fire and Police Station, Sheffield.
A large collection of old fire appliances, uniforms and photographs housed in one of the first purpose-built fire stations. Appointment necessary.